PLANT BIOLOGY RESEARCH AND TRAINING FOR THE 21ST CENTURY

Committee on an Examination of Plant-
Science Research Programs in the United States

Commission on Life Sciences

National Research Council

NATIONAL ACADEMY PRESS
Washington, D.C. 1992

This study by the Commission on Life Sciences was funded by the National Science Foundation, the Department of Agriculture, and the Department of Energy under contracts BBS-9007979 and DE-FG05-90ER 20004.

International Standard Book No. 0-309-04679-3
Library of Congress Catalog Card No. 91-67590

Additional copies of this report are available from:

National Academy Press
2101 Constitution Avenue, N.W.
Washington, DC 20418

S492

Printed in the United States of America

First Printing, June 1992
Second Printing, December 1992

iv

Preface

Plant research in the past hundred years has made major contributions to our understanding of biology. The earliest research in genetics developed from work on plants; molecular biology in its earliest days benefited from insights gained from studies on plant enzymes, viruses, and cytology. In the middle years of the twentieth century, from the 1930s through the 1960s and into the 1970s, special excitement in biology came from the elucidation of the physical and chemical basis of photosynthesis. Work with higher plants was critical to all those advances. Similar work with microorganisms began only after the foundation was laid by work in plant biology.

Contemporary biology derives its special excitement from the application of recent developments, not in plant biology, but from the microbial genetics work of the 1950s and 1960s and from the basic biomedical research successes of the 1970s and the 1980s. The new tools and paradigms of this more recent era have enlivened, and are revolutionizing, contemporary plant biology. Examples include the use of recombinant DNA technologies to develop transgenic plants and the study of the genetic basis of such phenomena as plant development, plant-microbe interactions, and plant reproductive biology.

Fundamental research on plants in earlier generations was critical to the development of biology and yielded important benefits to society. The intellectual excitement, productivity, and breadth that characterize the forefront of contemporary plant biology hold as great a promise. The plant-science community in academia, although diverse and often fragmented, is committed, imaginative, persistent, and resilient. But the potential scientific and tangible benefits to society of today's and tomorrow's research opportunities could remain unrealized. Relative to other scientific fields, and in proportion to its own shrinking numbers, the plant-biology research and training capacity of the United States seem to many in the field to be depleted. The data reviewed by the Committee on Plant

Sciences bear out that conclusion. The number of well-funded laboratories working with plant systems is small and the number of students at all levels, the number of course offerings, and the sense one gets of the stature of research on plants among scientists and the public seems to have declined in recent decades. Allocations for competitively awarded grants to fund research and training in the plant sciences are tiny in comparison with the life sciences as a whole. And major obstacles, of funding and of valuation by colleagues and administrators, face those who wish to integrate research and teaching about plants into the curriculum and into the fabric of the biologic sciences as a whole on many campuses.

Our committee was assembled in response to a request from the National Science Foundation (NSF), the U.S. Department of Agriculture (USDA), and the U.S. Department of Energy (DoE). The leadership of these agencies asked the National Academy of Sciences through the National Research Council (NRC) to assess the status of plant-science research in the United States in light of the opportunities arising from advances in other areas of biology. NRC was asked to suggest ways of accelerating the application of these new biologic concepts and tools to research in plant science with the aim of enhancing the acquisition of new knowledge about plants.

The committee was established in the Commission on Life Sciences in the fall of 1990 to conduct this assessment and to prepare appropriate recommendations. The charge to the committee was to examine the following:

- Organizations, departments, and institutions conducting plant biology research.
- Human resources involved in plant biology research.
- Graduate training programs in plant biology.
- Federal, state, and private sources of support for plant-biology research.
- The role of industry in conducting and supporting plant-biology research.

- The international status of U.S. plant-biology research.
- The relationship of plant biology to leading-edge research in biology.

The committee also was asked to recommend improvements in institutional and infrastructural arrangements that would enable plant scientists to function at the forefront of biologic research as they address scientific questions about agriculture and the environment.

After a discussion at its first meeting with Mary Clutter of NSF, Charles E. Hess of USDA, and Robert Rabson of DoE, the committee took it as its charge to consider the plant sciences in their broadest sense–the study of plants, at all levels of organization–as part of the search of new understanding and the elucidation of fundamental biologic principles. Our report focuses on research in basic plant biology and suggests changes to enable plant studies to function in the United States at the forefront of research, as have research on microorganisms and on animals for the past four decades.

Our report deals with the central role of plant biology and plant biologists in enabling the United States to meet challenges not only in agriculture but also in other applications of plant science. Knowledge gained from the study of plants has immediate applications to a wide range of problems and opportunities facing modern society. Not only is increased knowledge about plants fundamental to advances in agriculture and forestry, but it can contribute to advances in nutrition, to improved understanding of the environment and mitigation of global change, to the development of alternative sources of energy, to the development of manned space exploration, and to the production of improved medicine. The committee's report thus touches the interests not only of its sponsors–NSF, USDA, and DoE–but those of other agencies, such as the National Institutes of Health, the Environmental Protection Agency, and the National Aeronautics and Space Administration.

The committee met twice to discuss the issues and to plan, outline, and draft its report. Our time constraint was acute. Because of this limitation, the committee was unable to undertake extensive new research or analysis. Instead, we relied on NRC reports and on data from various federal agencies. The committee used reports from NSF, DoE, and USDA and relied on individual members' knowledge and experience as a basis for understanding the organization and structure of research in the plant sciences and the opportunities and needs therein. Also because of time and resource limitations, the committee was unable to address the international-status and private-sector aspects of plant biology.

We must thank many people for their generous sharing of time and expertise. Paul Stumpf, Jane Smith, and Patricia Shelton of the USDA National Research Initiatives Competitive Grants Office; Marge Stanton of the USDA Higher Education Office; Clifford Gabriel of the USDA Cooperative State Research Service; Judith Greenberg of the National Institute of General Medical Sciences; Machi Dilworth and Gerald Selzer of NSF; and Thora Halstead of the National Aeronautics and Space Administration supplied valuable information, opinions, and data on plant-science research programs. Neal Jorgensen, acting dean of the College of Agricultural and Life Sciences of the University of Wisconsin–Madison, helped the committee to understand issues in plant-science research and education in the university milieu. Elaine Hoagland of the Association of Systematics Collections graciously provided data on extant collections.

The statements and interpretations in this report, however, are the responsibility of the committee rather than of the persons who so kindly provided us with information.

I wish to extend special thanks to the members of the committee, who so enthusiastically, thoughtfully, and patiently applied themselves to this important task, and the members of the committee join me in thanking Commission on Life

Sciences staff members Alvin Lazen and Juliette Walker, and Board on Agriculture staff member James Tavares for their splendid assistance throughout the preparation of this report.

Robert M. Goodman
Chairman, Committee on an
Examination of Plant-Science
Research Programs in the United States

Contents

Over the past 25 years a major revolution has occurred in biology. Research advances, especially at the molecular level, have permitted exponential increases in our understanding of fundamental life processes. The leading edge of these advances has been in the biomedical disciplines. Human health-related areas have been the major beneficiaries.

There is an increasing realization that other potential beneficiaries of the biological revolution are the agricultural and environmental disciplines. Research opportunities abound in the plant sciences that could make a major impact. Yet, recent reports indicate that plant science has not kept pace with the forefront of biological research. It is time to address this disparity.
(Mary Clutter (NSF) letter to Frank Press, Oct. 25, 1989)

EXECUTIVE SUMMARY

Modern civilization rests on the successful and sustained cultivation of plants and on the wise use of the biologic and physical resource base on which their cultivation depends. Our knowledge about the world around us is incomplete if we do not include plants in our discoveries, and it is distorted if we do not place sufficient emphasis on plant life. From fundamental discoveries about plant life arise technologies and capabilities that have a wide range of practical applications.

Only higher plants and a few microorganisms can convert light energy from the sun into chemical energy. Photosynthetic organisms are at the center of the earth's hospitality to other life forms, and higher plants are important in regulating the earth's systems of atmosphere, water, and climate. We will never fully understand the global environment–or have a serious hope of successfully managing it in the face of explosive population growth, severe shifts in land use, and other effects of human habitation–until we have a much more comprehensive understanding of plants, their cellular processes, and their ecology and population biology.

Plants are critical to human health. They are the dietary source of essential amino acids, vitamins, and other nutrients. They are an important and original (and in many cases,

1

continuing) source of therapeutic drugs—more than 20% of all prescription drugs are derived from plants and many more were first discovered and formulated as plant products. The health of the human race could well rest on the quality and extent of our understanding of plants, their uses, and their requirements.

Examples from the past—from Mendel's discovery of the rules of genetic inheritance to the X-ray diffraction of tobacco mosaic virus, which paved the way to elucidating the structure of DNA—illustrate the importance of plants to biologic research. But how well equipped are we to deal with the opportunities and challenges that lie ahead?

The concerns that led to this study were that research in and teaching of the biology of plants have been insufficiently emphasized and that plant biology has become isolated from the mainstream of biology. The Committee on an Examination of Plant-Science Programs in the United States was established in the Commission on Life Sciences of the National Research Council to review the data available and to consider whether the academic and research institutions of this country are prepared to address the opportunities in modern plant biology. The committee also was asked to recommend how the nation might change its approach to the support of plant sciences to reduce the imbalance in the emphasis given in laboratories and classrooms to plant biology relative to other fields of biology.

This report focuses on three issues facing the plant sciences in academic research and training. First is the *mechanism* of research funding (competitive versus noncompetitive). Second is the *balance* of research funding (support of basic research into the nature of life processes versus applied or adaptive research). Third is the *commitment* to building and maintaining an appropriate infrastructure of institutions and personnel.

The members of this committee are convinced that the U.S. research effort in plant biology is not keeping pace with biomedically related fields because of the defective *mechanisms* used for support as well as the small financial commitment to research and training in plant sciences. The patchwork system of support for research and the incomplete system of support for career training in the plant sciences described in Chapter 2 creates impediments to the success of plant-biology research in the United States. These impediments include an insufficient focus on plant science as a basic discipline of biology; the isolation of plant sciences from other disciplines of biology; the insufficient funding and fragmentation of support for basic plant-biology research; the sometimes inappropriate philosophy and rationale for funding; and the insufficient support for training, instrumentation, and facilities. A downward spiral (Figure 1) has resulted from the

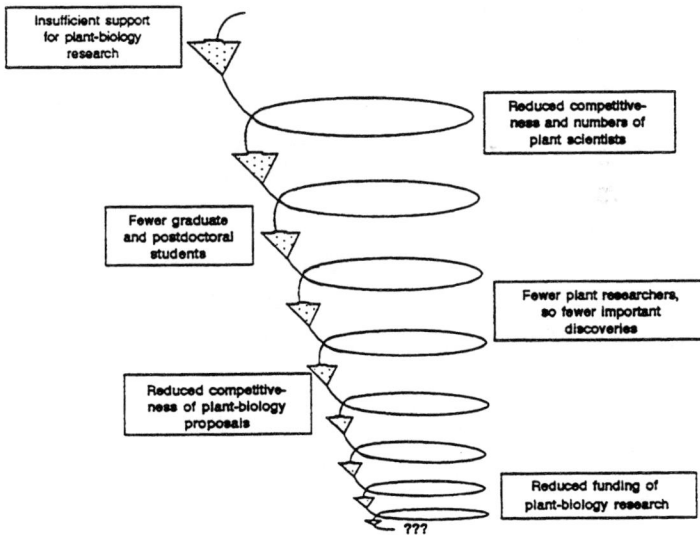

Figure 1   Downward spiral of basic plant-biology research and training.

lack of a coordinated program to support research, graduate training, and postdoctoral training and from the failure to establish linkages with other basic and applied disciplines.

In its broad outline, the remedy that the committee proposes is simple and proven. In the 45 years since the beginning of large-scale federal support of science, the strategies used by the various federal agencies to fund scientific research in support of societal goals have constituted an experiment. NIH and NSF have based funding decisions on competitive procedures designed to recognize individual merit; USDA has based funding decisions on institutional, political, and historical considerations that do not preclude but that also do not necessarily reward or reinforce individual merit. The committee concludes that the results of the experiment are clear. The philosophy, mechanisms, and strategy used by NIH and NSF to support basic research and its applications have advanced science of the highest quality, attracted the best young scientists to careers in research and teaching, and provided a stream of discoveries that has been rapid and highly beneficial to society. The success of the NIH and NSF grant programs has engendered their enthusiastic and generous support by Congress and successive administrations.

The committee's members believe it is time to take these lessons and apply them to the plant sciences. To this end, the committee recommends the establishment of a comprehensive program that engages all of the federal agencies that support plant biology. The recommended program would include the following components:

- Investigator-initiated competitive grants.
- Postdoctoral training.
- Predoctoral training (training grants and fellowships).
- Undergraduate training.
- Career training and redevelopment.

- Facility support.
- Meeting support.

The example in government closest to the philosophy and practices the committee recommends for federal support of plant science is the National Institute of General Medical Sciences (NIGMS) of the National Institutes of Health. NIGMS is the extramural arm of a federal agency with an applied mission and focuses on basic science and support of the scientific infrastructure in performing its mission.

The success of the proposed plant biology program will depend on its meeting the following criteria:

- The program should be dedicated to the study of plant biology as a basic science. It should not be a mission-oriented program aimed at solving specific practical problems.
- The program should encompass a comprehensive system of extramural research and training to include pre- and postdoctoral fellowships, training grants for graduate students, grants for the purchase and upkeep of instrumentation, and financial support for meetings. The system of grants should support the highest quality research in nonprofit institutions.
- The program should be patterned after the philosophy of the NIGMS.
- The program should support high-quality research being done by plant biologists in nonprofit institutions. Communication between plant scientists and researchers in other disciplines should be encouraged.
- The program should provide grants and fellowships in sufficient number and amount of award to attract and retain the best scientists.
- The program should be administered by an agency committed to the above standards.

The scope of the program should be carefully defined in the course of further study but the following subjects are cited as examples:

- *Subcellular processes*, including biochemistry, photochemistry, organelle structure and function, gene and chromosome structure, genome organization, mutagenesis and DNA repair, and gene expression and regulation.
- *Cellular processes*, including developmental biology and developmental genetics, signal transduction, cell-to-cell communication, cell division and growth, photosynthesis, and intercellular transport of water and nutrients.
- *Organismal processes*, including growth and reproductive biology, structure and function of plant organs, responses to the environment at the supercellular level, and nutrient and water transport in the whole plant.
- *Population and species processes*, including areas such as ecology, population biology and genetics, systematics, and issues of biodiversity.
- *Plant interactions with the biotic and abiotic environment*, including nitrogen fixation, interactions with beneficial microorganisms, pathogenesis, the genetics and molecular biology of plant defense and stress responses, and community ecology.

The committee presents in Chapter 4 three recommendations that embody the above principles and views.

RECOMMENDATION 1

A National Institute of Plant Biology (NIPB) should be established in the U.S. Department of Agriculture (USDA) under the direct oversight of the assistant secretary of agriculture for science and education.

**NIPB should be responsible for leading a coordinated
federal plant-biology program that intimately involves
other federal agencies that support research and
training in plant biology.**

Because plant biology encompasses far more than
agriculture and its applications, the historical mission of USDA
is too narrow to encompass the breadth of fundamental plant-
biology research and teaching as we envision it. Moreover, the
USDA has only recently, and very slowly, moved to adopt a
significant extramural component to its mission through a
program funding competitive grants. This program, begun in
1978, was an important change from almost exclusive concentration on formula funding by USDA. The fiscal year 1992
National Research Initiative Competitive Grants Program
(NRICGP) enlarges USDA's small program of grants to support
extramural plant-biology research related to agriculture. At full
funding it is planned that $125 million of the $500 million total
would be for plant biology. Building on the foundation of
NRICGP, we propose that the plant systems part of NRICGP
become the core of NIPB, which would serve as a primary focus
for research and training in the study of plant biology oriented
toward agriculture, food, and the environment. In addition,
NIPB would be the lead agency in coordinating the efforts of
other agencies in plant biology.

If USDA should prove unwilling to fulfill the role we have
proposed for it, NSF should be assigned the task of leading the
program. NSF has clearly demonstrated its dedication to the
support of fundamental research based on competitively
awarded, investigator-initiated grants.

Implementation of our proposal would require that USDA
effect major changes in its funding philosophy, its operational
patterns, and its relationship to Congress and the scientific
community. It will need to

- Plan beyond the design drawn for NRICGP and its proposed five-year funding strategy.
- Support evolution of NRICGP and its *competitive* grants program.
- Focus on the support of *fundamental* plant biology.
- Protect the new institute from political and commercial pressures.
- Avoid over managing the scientific research process.
- Demonstrate increased leadership in coordinating its work with that of other agencies.
- Develop department-supported training programs and encourage training programs at other agencies.
- Increase the use of peer review procedures that employs the expertise of the entire scientific community and reaches outside government agencies.
- Organize NIPB to ensure its high visibility, stature, and independence within the federal government.

The establishment by USDA of the institute would be another step in an important progression. The first step was the establishment of a competitive grants program; the second was the expansion of that program under the National Research Initiative. The next step is the expansion of the plant systems part of NRICGP to a national institute. Potentially, other parts of NRICGP, for example, the animal health program, could become institutes. Eventually, USDA could resemble the model of the Department of Health and Human Services and its National Institutes of Health for support of the extramural and intramural research, training, and infrastructural elements of sciences relevant to its mission.

The committee recognizes that the recommendation for the establishment of an institute is proposed at a time of both national budgetary constraint and while the USDA National Research Initiative is in mid-course of implementation. However, the increases in funds we have proposed for support of the NIPB at USDA and to increase expenditures at the other agencies that are major supporters of plant biology research and training are modest (see Chapter 5 for detail) amounting to about an additional $240 million annually by the year 2000. We believe further that the formation of NIPB is the next and natural step in the growth of the competitive grants program at USDA and needs to be discussed now, and the groundwork laid, before the completion of the five-year plan to build the National Research Initiative is completed.

This Report and *Investing in Research*

Concerns about a possible deficiency in knowledge about plants and inadequacies of research funding and manpower have been raised by others. Many of the concerns have centered around the need to solve urgent problems, such as global climate change, food shortages, undesirable consequences of some agricultural production methods, and loss of biologic diversity. Studies include one prepared by the National Research Council's Board on Agriculture. The report, *Investing in Research*, called for a major new initiative for agricultural research. *Investing in Research* has led to the incorporation of a National Research Initiative (NRI) into the Administration's FY 1991 and 1992 budgets for the U.S. Department of Agriculture. The major recommendation of *Investing in Research* was that USDA be authorized to enlarge, in both scope and funding, its competitive research-grants program. The focus of the NRI is agricultural, and it includes a program of research on plants.

The present report goes beyond *Investing in Research* in proposing changes in how plant biology is managed within the USDA and in a context broader than agriculture, and it contains recommendations about the USDA leadership responsibility for the health and strength of the research and research personnel infrastructure. We believe our proposals are consistent with the spirit of *Investing in Research* and are a necessary condition for the long-term success of the National Research Initiative.

RECOMMENDATION 2

All agencies that currently support plant-biology research and training should maintain and increase their commitment in cooperation with NIPB and USDA.

The National Science Foundation, the Department of Energy, the National Institutes of Health, and the National Aeronautics and Space Administration have provided valuable support for plant-biology research, and their continued–and increased–commitment is needed to fulfill the new institute's objectives. These agencies, together with USDA, encompass virtually all aspects of a comprehensive plant-biology program. The National Institutes of Health and the National Science Foundation could provide the training grants and fellowships that are essential for training more plant scientists. However, we urge USDA to consider developing large-scale training programs. For NIPB to succeed, all involved agencies must increase the amounts awarded for individual research grants. These agencies have demonstrated a remarkable degree of cooperation in the past, for example by making joint decisions for the funding of plant science centers. In September 1991, USDA, DoE, and NSF signed an agreement to continue their joint program on collaborative research in plant biology.

RECOMMENDATION 3

An independent group of nongovernment scientists should be formed to provide continuing advice to the USDA assistant secretary for science and education and to the officials of cooperating agencies concerning NIPB's operation and goals and to oversee the parallel efforts by other agencies.

In five years an independent group should examine and evaluate the progress of all of the agencies involved in implementing the recommendations contained in this report.

We believe special provision should be made to provide continuing, independent advice and to ensure that the program's effectiveness is evaluated after an appropriate period.

Chapter 4 of the report details the components of the comprehensive program—the minimum necessary to ensure U.S. leadership in plant-biology research into the next century. The estimated cost for the first year is $280 million; this includes about $150 million already allotted for competitive grants programs by several federal agencies. Our best cost estimate for the year 2000 is about $520 million (in 1991 dollars); this represents growth of about 6% annually. These sums are modest considering the contribution plant biology research can make to maintaining the U.S. role as a global leader in agriculture, the environment, health and medicine, and science education.

1

WHY PLANT-BIOLOGY RESEARCH TODAY?

Throughout human history, plants have been the object of pervasive and at times dominant artistic and intellectual interest. Plants were important subjects from the earliest study of life processes, and they were central to scientific study in the nineteenth and early twentieth centuries.

Good reasons remain to study the basic life processes of plants. Research on plants enriches our intellectual life and adds to our knowledge about other life processes. The results of research on plant systems also can teach us *how* to approach problems in agriculture, health, and the environment.

PLANTS, HUMAN HEALTH, AND CIVILIZATION

Our understanding of plant life underpins a vast range of activities and touches virtually every aspect of human life. From their origins, human civilizations have depended for their development and prosperity on their ability to manage plants and have sometimes fallen because of their failure to do so. Throughout history, plants have been collected, traded, selectively adapted for new environments, and bred for new combinations of traits. Plants have been manipulated for use as food and fiber, and for many other, particularly aesthetic, purposes.

Modern civilization rests on the successful and sustained cultivation of plants and on the wise use of the biologic and physical resource base on which their cultivation depends. Our knowledge about the world around us is incomplete if we do not include plants in our discoveries, and it is distorted if we do not place sufficient emphasis on plant life. There are many compelling practical reasons also for society to invest in research about plants and to educate its citizens for careers in which knowledge about plants is important. From fundamental discoveries about plant life arise technologies and capabilities in a wide range of practical applications (Figure 2).

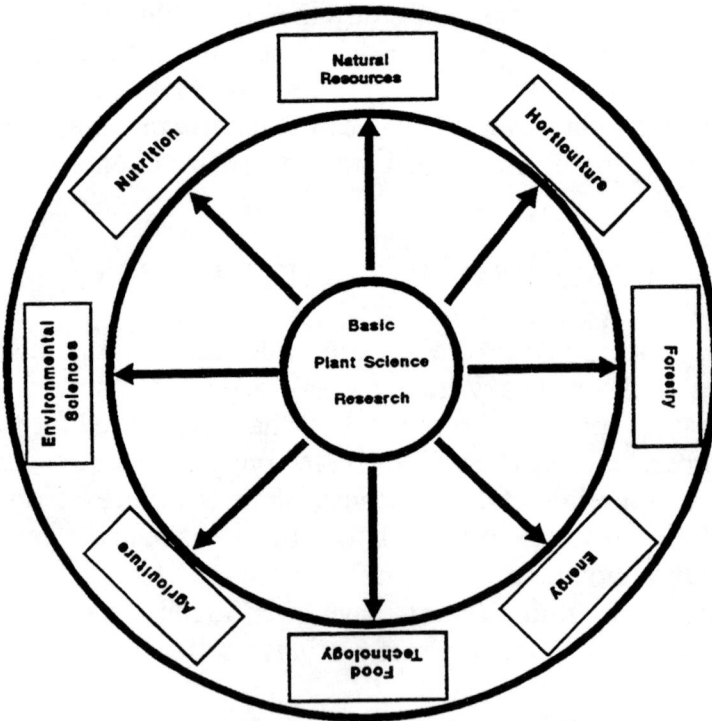

Figure 2   Potential applications of plant-biology research.

PLANTS AND THE ENVIRONMENT

Only higher plants and a few microorganisms can convert light energy from the sun into chemical energy. Photosynthetic organisms are at the center of the earth's hospitality to other life. Plants and photosynthetic bacteria gave rise to the earth's atmosphere. They are important in regulating climate and the chemical and biologic conditions of the soil and water. Photosynthetic plants are the source of the fossil fuels we are depleting today, and they provide the most readily harvested source of renewable energy for tomorrow. The primary atmospheric gas incorporated by plants in photosynthesis, carbon dioxide, is one of the major "greenhouse" gases. Plants regulate the carbon cycle of the biosphere. Plants, in part through their unique symbiotic relationships with microorganisms, also play a major role in regulating the partitioning of nitrogen between atmospheric and life processes. We will never fully understand the global environment–or have a serious hope of successfully managing it in the face of explosive population growth–until we have a much more comprehensive understanding of plants, their cellular processes, and their ecology and population biology.

Plants are important in maintaining a healthy environment, for example, by controlling erosion and water pollution, and by helping to reduce air pollution. They improve the environment for human activities everywhere–from indoor spaces to vast wilderness areas.

The role of terrestrial plants and marine phytoplankton in maintaining an environment suitable for human habitation is inadequately appreciated, but there is a growing recognition of the urgent need to illuminate the role of plants. The accumulated effects of more than a century of industrial activity, explosive population growth, severe shifts in land use,

Unique and Scientifically
Interesting Properties of Plants

Plants differ from animals in several important ways.

Development. The growth of a plant from an undifferentiated cell into a complete and mature organism requires only a few hormones. Moreover, plant cells are totipotent: It is possible to regenerate a whole plant from a single leaf or root cell. In contrast, specific cells (the germ line) of an animal in early development form the germ cells. Plants have no germ line in this sense and produce sexual organs and gametes from somatic tissue late in their development.

Biochemistry. Plants are virtually the sole source of new oxygen and carbohydrates on the planet. Light is harvested by unique organelles, the chloroplasts. Plants synthesize the 20 amino acids required for proteins, including the 10 amino acids that humans are unable to produce. Moreover, in a unique symbiotic relationship with some plants, microorganisms can fix atmospheric nitrogen for plant use in the synthesis of amino acids, proteins, and other compounds.

Physiology. Plants lack the major organ systems present in animals. Yet, their physiology permits them to respond to their environment. Instead of an immune system, they have inducible disease resistance mechanisms that enable them to make natural toxins against fungal and bacterial pathogens. Instead of a nervous system, they have a repertoire of receptors and pigments that allow them to respond to their environment. Instead of a muscular and skeletal system, they have a novel set of fibers for support. They are attached to their substrates, and they can move only by growing or by gaining or losing water.

Plants and Global Warming

Atmospheric modelers are trying to evaluate the effects of changes in carbon dioxide concentration on global weather patterns and temperature. Models that predict carbon dioxide uptake and water loss by leaves grown under different environmental conditions can make an important contribution to elucidating global climate change. Other plant research is needed to develop sensitive ways to determine how much of the light energy absorbed by a leaf is used for photosynthesis (for metabolism and growth) and how much is simply reradiated as heat. The efficiency with which plants use light can vary enormously in response to environmental variables, such as water stress, temperature, disease or insect damage, or fluctuations in the supply of nitrogen or phosphorus. Theoretical models are being rigorously tested, with a fair degree of success. In addition, remote-sensing techniques are being developed to evaluate the photosynthetic performance of whole plant communities in response to stress. Modeling and experimental studies promise the quantitative information required to put predictions of atmospheric change (or lack of it) on a sound basis.

and other effects of human use of the earth show that human activities can overpower the buffering effects of the natural processes that regulate global climate. The health and well-being of the human race could well rest on our achieving a better understanding on which to base a more reasoned exploitation of plant life.

PLANTS IN AGRICULTURE, MEDICINE, AND INDUSTRY

Macroscopic and microscopic plants form the first link in the terrestrial and aquatic food chains. Plants are thus at the heart of agriculture. Together with microorganisms and domesticated animals, plants provide the raw materials for our food and drink. Plants also provide many of the materials used in clothing and buildings. The application of basic knowledge about plants has made modern agriculture possible. For example, studies of the nutrient requirements of plants led to soil fertility management.

The Green Revolution was founded on fundamental knowledge gleaned from research in genetics and plant nutrition. Genetic manipulation is a powerful, proven method for improving the productivity, quality, and disease resistance of plants. Basic knowledge of genetic inheritance, defense responses, pathogen genetics, and population genetics will continue to yield improvements in the technology needed to secure a stable food supply.

Plants are critical to human health. They are the sole source of some of the essential amino acids, vitamins, and other nutrients in our diet. Research with plants was central to elucidating the role of vitamins in human health and disease: Plants high in ascorbic acid, such as peppers and citrus, prevent scurvy. Grains in the diet provide B vitamins. Many drugs were first discovered as plant products before methods for their synthesis were developed. Research on plants yielded cardiac glycosides (such as digitalis), a wide range of useful alkaloids (such as scopolamine, atropine, quinine, and ephedrine), dicoumarol, and many other drugs. Research on lower plants and agricultural soils yielded many antibiotics. Even today, more than 20 percent of all prescription drugs are derived from plants.

The chemical industry developed from the work of German scientists who learned to synthesize dyes from coal tar, a derivative of fossil plants, to replace the commonly used dyes derived from wild and cultivated plants. Now, the search has been reversed and plant-derived products are sought to replace harmful coal tar dyes. Modern industry and society continue to depend in many ways on chemical products derived from plants, such as soaps, detergents, rubber, paints, resins, plastics, adsorbents, and adhesives.

PLANTS AND THE ORIGINS OF MODERN BIOLOGY

Research with plants has strongly influenced the development of biology and has contributed to many important scientific advances. It was research with plants that led to the discovery of the rules of genetic inheritance (Gregor Mendel's peas), of the role of light in regulating the physiologic responses of higher organisms (phytochromes), of transposition of genetic elements (controlling elements in maize), and of the protein nature of enzymes (urease). Research with a plant virus contributed to the elucidation of the structure of DNA itself (X-ray diffraction with tobacco mosaic virus) and of the role of nucleic acids in the genetic material of all life forms.

These examples illustrate how the study of plants has affected biologic research for several generations. But how well equipped are we to deal with the opportunities and challenges that lie ahead? The techniques of modern biology, and in particular modern genetics, make many difficult problems in plant biology approachable. Before the era of recombinant DNA, the tools available for genetic studies of plants' development, metabolism, and environmental responsiveness were relatively crude. Now modern genetics offers new promise to the plant sciences. In some fields of modern biology, plants offer the preferred model system for fundamental and exploratory science through application of molecular genetic techniques. Scientists now can transfer genes easily among plant species, and because the genomes of some plant species are quite small they can be studied readily. Plants can be used to answer many general questions in biology in such diverse subdisciplines as development, metabolism, gene regulation, symbiosis, and chromosome structure.

It is not within the scope of this report to describe a research agenda for plant sciences. Other National Research Council reports have contained pertinent research agendas, for example, *Investing in Research* (NRC, 1989a), *Opportunities in Biology* (NRC, 1989b), and *Forestry Research: A Mandate for Change* (NRC, 1990).

In recent years, the scientific community has shown significantly increased interest in research with plants. The power of modern methods to answer important questions in plant biology has stimulated the interest of scientists in leading universities and other research institutions in the United States and abroad. Well-funded plant-biology laboratories here and elsewhere are making research contributions at the cutting edge of biology. This heightened interest has generated more worthy research proposals than public agencies can fund. An informal survey of the private sector in agricultural biotechnology indicates that in the late 1980s about $250 million (exclusive of development costs) each year was being spent on basic plant-biology research by companies whose work was primarily or exclusively with plants.

The fertility of modern plant-biology research is demonstrated in special issues of *Science* (November 16, 1990) and *Cell* (January 27, 1989). Developmental biology, cell-to-cell recognition, signal transduction, the molecular basis of disease, plant-microbe interactions, gene regulation, transposition, and photosynthesis are some of the areas covered in these issues. Several new plant journals have been launched recently; three leading examples are: *The Plant Cell, The Plant Journal,* and *Plant Molecular Biology.*

2

FEDERAL MANAGEMENT AND SUPPORT

Federal support of plant-science research in the United States now comes chiefly from the U.S. Department of Agriculture (USDA), the National Science Foundation (NSF), the Department of Energy (DoE), and the National Institutes of Health (NIH). Other agencies that provide lesser support are the Department of the Interior, the National Aeronautics and Space Administration, the National Oceanic and Atmospheric Administration, and the Office of Naval Research.

The information in this section of the report is taken from a variety of public sources. Much of the data and inferences we present are based on reports from NSF (NSF, 1990b) and the American Association for the Advancement of Science (AAAS, 1990). Analysis and comparisons are difficult because the data were generated for a variety of purposes. Agencies report expenditures and other data with different definitions of disciplines and without agreement about whether specific research programs are "basic" or "applied" and whether grants are "competitive" or "noncompetitive." Thus, the expenditure figures in this report are best estimates based on the committee's interpretations. The data describe a patchwork of funding for plant research in the United States from five federal agencies with different policies and practices.

THE FEDERAL INSTITUTIONS

U.S. Department of Agriculture

The research programs of USDA began in the land grant colleges with the signing of the Morrill Act in 1862. (A history of the origins and provisions of the formula grant program can be found in Kerr, 1987.) In 1887 the Hatch act provided annual funding to support state agricultural experiment stations. The Smith-Lever Act of 1914 established cooperative extension programs at the land grant colleges. In 1962, the McIntire-Stennis Act gave funding to public colleges and universities for forestry research and graduate programs.

USDA agencies that conduct a significant amount of plant-biology research include the Agricultural Research Service (ARS), the Cooperative State Research Service (CSRS), and the Forest Service (FS). ARS and CSRS are under the Assistant Secretary for Science and Education and FS is under the Assistant Secretary for Natural Resources and the Environment. CSRS supports research scientists primarily associated with land-grant college and university agriculture experiment stations (AES). ARS and FS have intramural research programs in agriculture and forestry, respectively. USDA intramural research is also performed at research centers and by scientists located at universities.

The primary focus of this analysis of USDA plant-biology research is on CSRS and its three principal mechanisms of support. These are formula funding to State Agricultural Experiment Stations associated with land-grant colleges and universities, Special Research Grants that are either Congressionally earmarked to specific research programs or are awarded competitively, and competitive grants.

Formula funding is commonly referred to as base support for agriculture experiment station scientists and is spent largely at the discretion of individual AES directors. The majority of the funds are used for salary support.

The CSRS competitive grants program under the National Research Initiative supports peer-reviewed, investigator-initiated grants in six major categories, two of which, plant systems and natural resources and the environment, are directly relevant to plant-biology research. When the NRI is fully funded, $250 million annually will be devoted to supporting grants in these two categories.

The distribution of funds among these programs is shown in Table 1.

Table 1   Actual and estimated USDA expenditures for research and development, 1989 to 1991, for selected areas, in millions of dollars

	FY 1989 Actual	FY 1990 Estimate	FY 1991 Budget
Agricultural Research Service	525.3	546.9	587.8
Cooperative State Research Service	307.9	324.9	335.4
Competitive research grants	39.7	42.5	100.0
Agricultural experiment stations	155.5	155.1	158.5
Cooperative forestry research	17.5	17.3	13.0
1890 colleges and Tuskegee	24.3	27.7	31.5
Special research grants	45.6	56.3	25.6
Alternative crops research	1.0	0.3	0.9
Agriculture productivity	4.5	4.4	4.5
Other	19.8	21.3	1.4
Forest Service	138.3	148.0	160.0

Source: Excerpted from Table II-15, AAAS, 1990.

Funding for Plant-Science Research

USDA provides by far the largest amount of federal funding for plant-science research, and the scientific questions addressed are justified primarily by their applicability to the production and processing of agricultural and forest products.

In response to our inquiry, USDA's Current Research Information Service provided data on funding for plant research. In 1988, USDA allocated $300 million for its intramural program of research in plant sciences administered by the Agricultural Research Service. About $100 million was allocated through CSRS; $70 million was awarded by other USDA units for research on plants at AES and other cooperating institutions. Under the provisions of the Hatch Act and the McIntire-Stennis Act that govern CSRS allocations to the states, state governments must match (often by multiples of twice or more depending on the specific funding authority) the formula funds received from the federal government. Total formula-based funding far exceeds the federal contribution because of this state participation in plant-science research.

Of the total USDA allocation reported to the committee, 94% is non-competitively awarded to land grant colleges for support of intramural research by Agricultural Research Service (ARS) employees and scientists at state agricultural experiment stations, and for special grants, often awarded at the direction of Congress. Some of the institutions that receive formula funds use a system of intramural peer review of investigator-initiated proposals as a basis for distributing the funds, but local peer review seldom is as rigorous as is peer review by NIH, NSF, or the NRICGP.

Support for Training

USDA formula funds provided to the states can be used for training as well as research and many research assistants are supported by funds received by their supervisors. ARS has a program that supports about 20 postdoctoral fellows. However, USDA and other federal programs explicitly designed to train the next generation of scientists for careers directed to agriculture, food, and the environment are relatively new and only modestly supported.

The Food and Agricultural Sciences National Needs Graduate Fellowships Grant Program, which began in 1984 by supporting 302 predoctoral trainees, supported only 58 predoctoral fellows in 1989. Some grants were for plant research; for example, in plant biotechnology and forestry (USDA, 1990a). Many graduate students and perhaps a few postdoctoral fellows are supported by funds received by their supervisors from USDA under the formula grant system.

Competitively Awarded Research and Training Grants

A distinctive feature of CSRS funding of plant biology at academic institutions is that most grants are allocated to *selected institutions* by formula rather than through open competition among scientists in all research laboratories. The program of the Competitive Research Grants Office (CRGO, now NRICGP) originated in 1978 to award USDA funds competitively. The competitive-grants research program was a major departure from other USDA programs. *All* scientists at U.S. institutions working on science questions pertinent to a range of identified needs of U.S. agriculture are eligible to apply for funds. Its initial appropriation, in 1978, was $15 million; in 1990, it made awards of about $46 million, including about $27 million for projects in the plant sciences. Stimulated by the National Research Initiative (USDA, 1990b), NRICGP funding grew to $73 million in 1991 and is projected to grow to about $100 million in 1992. When the initiative is fully funded at a total of $500 million, it is projected to budget $125 million for research in plant systems. Another $125 million is scheduled for the plant-biology related program in natural resources and the environment.

Table 2 and Figure 3 show competitively-awarded funds for plant-biology research from federal agencies.

Table 2   Federal competitive grant awards for plant-science research,
           1990 [a]

Agency	Awards
Department of Agriculture Competitive Research Grants Office (excluding animal research)	$26,978,318
Department of Energy Division of Energy Biosciences	18,668,092
National Aeronautics and Space Administration Division of Life Sciences	3,200,000
National Institutes of Health National Institute of General Medical Sciences [b]	12,500,000
National Science Foundation, Directorate for Biological, Behavioral and Social Sciences (BBS)	69,854,198
Division of Molecular Biosciences	15,790,736
Division of Cellular Biosciences	19,178,209
Division of Instrumentation and Resources	
Special Programs of BBS	2,764,789
Instrumentation and Instrumentation Development	496,508
Division of Biotic Systems and Resources	
Ecosystem Studies [c]	17,029,316
Biological Research Resources [d]	1,189,585
Ecology Program	3,886,764
Systematics Program	5,569,181
Population Biology and Physiological Ecology	3,949,110
Total	$131,200,608

[a] Some awards made in fiscal year 1990 are multiyear awards.
[b] Estimate for research on higher plants.
[c] Estimate includes the entire program's expenditures.
[d] Collections used in support of plant-biology research.

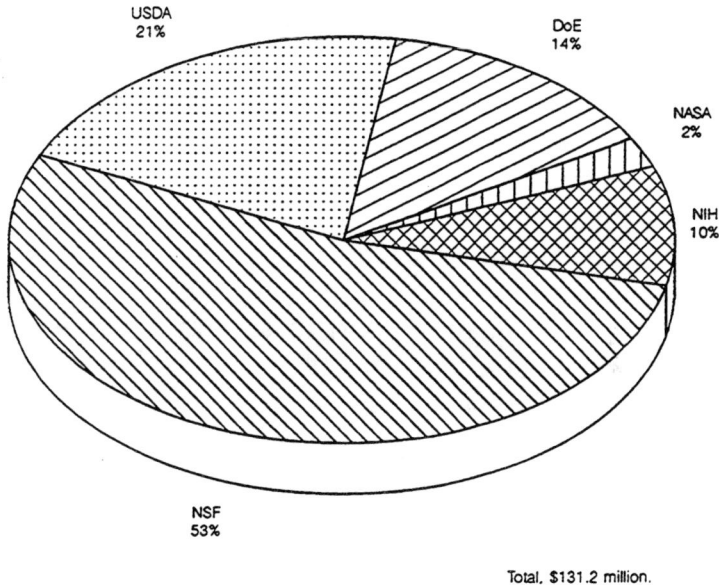

Figure 3 Percentage distribution among federal agencies of competitive awards funding for plant-science research, fiscal year 1990.

USDA 21%

DoE 14%

NASA 2%

NIH 10%

NSF 53%

Total, $131.2 million.

Department of Energy

The DoE Division of Energy Biosciences provided about $18.7 million in support of plant-science research in 1990 as part of its mission to explore biologic processes of potential use in energy production. The division uses a merit review system for decisions about awards of grants. After determining that a proposal meets basic standards of scientific merit, completeness, and compatibility with the DoE mission, agency project managers use DoE personnel and, in most cases, an ad hoc panel of external experts, to review the applications. DoE staff members interpret the reviews and discussions and make grant award decisions.

DoE-supported research is performed primarily in universities (80% of all funds and 87% of all grants awarded by the Division of Energy Biosciences in 1990) and national laboratories (11% of all funds) and it includes basic biologic studies of plant physiology, biochemistry, pathology, and genetics. Of the 172 projects funded by the Energy Biology Program in 1990, 150 were at universities and nonprofit institutions. Fifteen of that number went to the Michigan State University Plant Research Laboratory and two went to other University Plant Science Centers. Of the other 22 projects, 12 were at three national laboratories; the remaining 10 were at a variety of institutions.

DoE provides major support to several institutions. The Plant Research Laboratory at Michigan State University, which has been supported by DoE for many years and is subject to regular review, received $2.6 million in fiscal year 1990. The University of Georgia, including the Complex Carbohydrate Research Center, had about $3.5 million in grants in 1990 (many are for multiyear support). The Center for the Study of Early Events in Photosynthesis at Arizona State University received $1.2 million in 1988 to cover 30 months. (All of the above data are from DoE, 1990)

National Science Foundation

The NSF Directorate of Biological, Behavioral, and Social Sciences (BBS) provides about $70 million annually in support of plant-biology research. NSF is the largest federal provider of competitively awarded research grants in basic plant biology, and support for this area accounts for almost 25% of all funds awarded by BBS. Funding comes from the BBS Divisions of Molecular Biosciences, Cellular Biosciences, Instrumentation and Resources, and Biotic Systems and Resources and from specific programs within the divisions, which represent a wide array of disciplines, including ecology, systematics, and population biology.

NSF's review procedure invests significant decision-making authority in program managers, who obtain advice from a standing advisory committee and outside reviewers. Program managers' decisions are reviewed and approved by the heads of the divisions and the directorate.

NSF support for research is part of its extensive program of research grants, symposia, and other meetings, and its funding of instrumentation and resources. The foundation provides support for several large-scale projects: It received $4.4 million for 1991 in support of the *Arabidopsis* genome project, a major initiative in plant-biology research (NSF, 1990a). A Michigan State University science and technology center focusing on microbial ecology is supported by NSF. Another center, for developing strategies to engineer plants for resistance to pathogens, was founded in 1991 at the University of California, Davis. Center awards are made competitively for fixed terms and are subject to periodic review.

Included in this NSF support is more than $2 million for about 80 postdoctoral fellowships and approximately $500,000 for about 20 Presidential Young Investigators who work on basic plant biology. Some predoctoral fellowships in plant sciences are provided through the NSF Education Directorate. In late 1990, BBS awarded its first 10 training grants, including one from the BBS Research Training Groups Program for a plant-cell-biology program at the University of California, Davis. The training grants typically support five to 10 graduate students as well as undergraduates and postdoctoral fellows. Grants are about $1.5 million each for five years and can provide funds for instrumentation. These training grants are an important new initiative in NSF funding.

National Institutes of Health

In keeping with its focus on studies of basic biologic processes, NIH provided through the National Institute of General Medical Sciences (NIGMS) about $12.5 million in 1990 for research on higher plants through investigator-initiated, competitively awarded grants. We have not included in our summary about $45 million awarded by NIGMS for research on yeast models. All NIH grants and fellowships are awarded competitively on scientific merit, as judged by study sections of scientific peers.

Multiagency Cooperation

Federal organizations cooperate in support of research. For example, USDA, DoE, and NSF decide together on how to support centers such as the one for the study of complex carbohydrates at the University of Georgia and the center for the study of lipid and starch biosynthesis at Michigan State University. However, for administrative reasons, such centers are funded by one agency–for example, DoE, in the case of the University of Georgia, and USDA, in the instance of Michigan State University. An Interagency Plant Science Committee has representatives from USDA, NIH, DoE, NSF, the Department of the Interior, the National Aeronautics and Space Administration, and the Office of Naval Research; it meets regularly to discuss issues of common interest. The *Arabidopsis* Genome Research Project is coordinated under an interagency agreement among NSF, the NIH National Center for Human Genome Research, USDA, and DoE. In September 1991, USDA, DoE, and NSF signed an agreement to continue their joint program on collaborative research in plant biology.

IMPEDIMENTS TO AN EFFECTIVE PROGRAM OF BASIC PLANT-BIOLOGY RESEARCH AND TRAINING

The committee members have pooled their experience and compiled information obtained in interviews with other scientists in formulating the following list of impediments to plant-biology research and training. The committee has considered the issues within a framework that rests on three main principles:

- Basic research to elucidate fundamental processes often leads to unexpected results that can have great practical value.
- Science progresses best when the ideas for research are conceived and the work performed by researchers in individual laboratories rather than by highly managed teams or groups focused on applied research.
- An important responsibility of an academic community is education and training that will provide a steady stream of new investigators.

Lack of Focus on Plant Science as an Important Basic Biology Discipline

Most of the federal funds that support research in the plant sciences are allocated for programs that target practical problems rather than the understanding of basic plant processes. Biomedical research is directed toward practical questions, but NIH has emphasized *basic* biomedical research. That philosophy and NIH's generous funding have produced a continuing stream of discoveries in medicine, the development of a new biotechnology industry, and worldwide U.S. leadership in biomedical research.

Plant Sciences are Isolated from
Other Disciplines of Biology

Land grant universities have been the primary source of research and graduate education in the plant sciences. Of 724 Ph.D. recipients in plant biology in 1988-1989, 90% were from land grant universities (NSF, 1990b Table A 11). The pattern differs dramatically in biomedical research, in which one-third of the doctorates are awarded by private universities.

The absence of plant-science research and training programs from many of the institutions where health-related research has flourished has reduced the opportunities for communication between disciplines and worked to the disadvantage of both research areas. The divergence has become more pronounced with the rapid advance in knowledge of molecular processes. Plant biology programs often are isolated from other research and teaching in biology, even in broadly based and productive institutions. There are exceptions, and some schools, both public and private, have highly effective research and training in the full range of biologic systems, including prokaryotic, fungal, plant, and animal biology.

Funding for Basic Plant-Biology
Research is Insufficient

The $131.2 million in federal money spent by a variety of entities in fiscal year 1990 on competitively awarded grants for basic plant-biology research is small compared with the amount spent on many other federal scientific research programs. NIH institutes range in expenditures from $115 million for the new National Institute of Deafness and Communications Disorders to $1.6 billion for the National Cancer Institute. The institute median is about $450 million. About 80% of the expenditures are for extramural research. The budget of NIGMS (the institute closest in spirit to the model we propose for an institute of basic plant biology) was $629 million in 1990 (AAAS, 1990). In comparing basic plant biology with basic biomedicine, it is apparent that plant biology lacks a coherent program and adequate financial support for training and research.

Total federal support for basic research in 1990 was $11.4 billion; about $5.2 billion went for basic life-science research (NSF, 1990c). The $131.2 million of competitively awarded funds for basic plant biology was only 2.5% of the total expenditure for life-science research.

Funding is Awarded to Specific Institutions

USDA's research support is directed predominantly to land grant universities and to intramural research at ARS. Much of the support is awarded by formula to qualifying land grant institutions. The money is used effectively, but it is not available to investigators at other institutions, including those at which a substantial portion of the nation's high-quality biologic research is conducted. In addition, only those faculty at land grant institutions who have agricultural experiment station appointments have access to formula funds.

Grants for Plant Research are Smaller

The awards from peer-reviewed grant programs at NSF, DoE, and NRICGP are typically less than $100,000 per year for a term of two or three years. In comparison, the average grant from the National Institute of General Medical Sciences is $170,000 per year for four years (USDHHS, 1991). The short-term, small grants for plant research are usually just adequate to provide salary for perhaps two trainees plus a modest amount for laboratory operating expenses and overhead. As a consequence, the development of a plant-science research laboratory large enough to compete effectively in today's world of science requires that an investigator acquire several small grants and submit new proposals frequently. This constraint tends to focus research into small, short-term packages, often molded to match the missions of different granting agencies.

Financial Support for Training in Plant Sciences is Inadequate and Undependable

Funding for direct support of predoctoral and postdoctoral training in plant sciences is commonly inadequate and undependable. In 1984, $5 million was appropriated to establish the USDA National Needs Graduate Fellowship Program, which enabled 302 students to enroll in a wide range of graduate degree programs; funding ceased in 1986. In 1987, USDA provided $2.8 million for new fellowships, and this funding remained constant for fiscal year 1989. USDA has been unable to provide sufficient funding or to sustain the modest programs it has begun. ARS has a program in support of about 25 postdoctoral fellowships in ARS laboratories.

NSF's support of training is modest and the substantial NIH support is not directed primarily toward development of plant scientists. NSF (1990b) reported that there were 7,317 graduate students in plant biology disciplines and about 1,120 postdoctoral fellows in 1988-1989. Federal fellowships support 7% of the postdoctoral trainees and 4% of the graduate students, whereas federal research grants support 21% of the graduate students and 53% of the postdoctoral fellows. Graduate students also receive support from various other sources, including institutions (28%), state governments (15%), and personal funds (11%). Other sources of support for postdoctoral fellows include state governments (11%), foreign governments (7%), and industry (7%). Compared with a total of about 2,100 federally supported trainees in plant biology, NIH supported about 11,000 predoctoral and postdoctoral trainees in human health through fellowships and perhaps another 4,500 through research grants in 1989 (IOM, 1990).

Although the elements of a desirable training sequence exist in the form of fellowships and some support for training, they are unconnected–pipes, not a pipeline. The current system for the support of undergraduate, graduate, and postdoctoral training lacks the structure and continuity necessary for ready progression from one level to the next.

- At the undergraduate level, plant science often is missing from departments of biology. This is especially true in private universities; only 280 (6%) of a total of 4,517 full-time plant biology faculty were at private institutions in 1988-1989 (NSF, 1990b, Table A-3). Thus, many undergraduates are never exposed to plant biology.
- At the graduate level, eight of the 25 top-ranked universities in order of receipt of federal funds for life science research do not have doctoral programs in plant biology (NSF, 1990b).
- Many schools that do teach plant biology fail to provide adequate training for undergraduates who wish to pursue careers in plant research. Undergraduates often do not have the basic scientific training required to compete for entrance to the available graduate programs in basic biology.
- Insufficiently rigorous training of students has a negative effect on plant-biology research. In a recent NSF poll, 53% of the representatives of major academic programs in plant biology cited "poor quality of graduate and undergraduate students" as a factor that limited progress in plant biology (NSF, 1990b, Figure 11).

- Although NSF and NIH provide some postdoctoral awards for plant biologists, neither supports the substantial amount of graduate training in plant biology that is prerequisite for postdoctoral work.

The Need for Instrumentation and Facilities is Critical

A survey of instrumentation and facilities needs for agricultural biotechnology, conducted at 13 representative U.S. land grant universities and two private companies in 1989, has shown a critical need for instruments, modern laboratory space, and consumable materials (NASULGC, 1989). Although the study was intended to gauge the needs for agricultural biotechnology, we perceive its findings to mirror a pervasive problem in all segments of plant science. Collections and herbaria that serve as important research resources generally are poorly supported. There is a need to ensure that these resources are not lost through neglect.

The modern achievements of skill, enterprise, and science, new ideas with germs of power, must be recognized and diligently studied, as they have brought and will continue to bring daily competition which must be met.

If the world moves at ten knots an hour, those whose speed is but six will be left in the lurch.

(Congressman Justin Smith Morrill in 1859, three years before passage of the Morrill Land Grant Act.)

3

FINDINGS AND CONCLUSIONS

The biology of plant life should be a central focus for the nation's investment in research and teaching. The years since World War II have seen an extraordinary development in biology and biomedicine, but the study of plants has lagged behind.

The past 45 years have marked an experiment in the public support of basic and applied biologic research. Support of biomedical applications has been based on competitively awarded, investigator-initiated grants. The comprehensive system has included training, career development, and institutional awards, as well as competitive, investigator-initiated grants. The system has been open to competition among all sectors of the nation's diverse research community.

Support of research on plants has been directed primarily to research targeted to applications in agriculture, food, and energy. Although the U.S. Department of Agriculture (USDA) provides substantial support for plant research, funding has been predominantly in the form of formula allocations to a fraction of the nation's research and teaching institutions. The USDA Agricultural Research Service also constitutes a relatively large federally supported intramural program, but this system is not comprehensive and it is not open. A comprehensive system includes training, career development, and institutional awards, as well as competitive, investigator-initiated grants. An open system allows competition for grants by all sectors of the nation's diverse research community. Competitively awarded, investigator-initiated grants are only a small portion of the overall support for plant science research. The National Science Foundation (NSF) and, until recently, the National Institutes of Health (NIH), have historically awarded more funds for plant research and training through competitive, investigator-initiated granting programs than has USDA.

The members of this committee were convinced that the U.S. research effort in plant biology is not keeping pace with work in biomedically related fields because of the inadequacy of the mechanisms and funding used to support plant-science research. We believe that the federal government needs to alter dramatically its management and support of plant biology.

There are three related issues on which this report focuses in suggesting remedies for the deterioration of the plant sciences in the academic research and training enterprise. First is the *mechanism* of research funding (competitive versus noncompetitive; open to the larger scientific community versus closed). Second is the *balance* of research funding (support of basic versus applied research). Third is the *commitment* to building and maintaining an appropriate infrastructure of institutions and personnel (the amount of funding for research and training of the next generation of plant scientists).

The stunting of plant sciences at a time when other fields are experiencing rapid growth initiates a self-perpetuating downward spiral in the plant sciences. As universities restructure traditional botany, zoology, and microbiology departments into thematic departments that cut across organismal boundaries, plant biology loses academic positions to fields with access to the much larger funding bases of the biomedical support structure. In 1988-1989, only 16% of plant-biology faculty were at universities that had demonstrated their competitiveness in science by ranking among the top 20 institutional recipients of federal support for research and development in the life sciences (NSF, 1990b). Between 1982-1983 and 1989-1990 the number of plant-biology faculty at the top 20 universities decreased from 4,607 to 4,517 (NSF, 1990b). In contrast, these same institutions have a substantial representation of biomedical scientists and their placement in the top 20 recipients of federal support for research derives primarily from funds received from NIH. This unequal allocation of research and training resources has induced the documented paucity of faculty, research, and training in plant biology throughout academia. Unless the number of plant scientists in college and university biology departments is raised, many undergraduate and graduate students will never be exposed to plant biology. Many introductory biology courses include little information on plants, and many advanced texts in cell or molecular biology minimize the discussion of plants. Few students will choose to enter a field whose apparent lack of importance is documented by its absence from courses and texts. It is not surprising that the enrollment in baccalaureate plant sciences programs, as reported by members of the National Association of State Universities and Land Grant Colleges, decreased from 10,953 in 1982 to 6,974 in 1989 (NASULGC, 1990) or that the number of plant-science graduate students shrank from 8,023 in 1982-1983 to 7,317 in 1988-1989 (NSF, 1990b, Table A-8). The number of baccalaureate degrees awarded in the life sciences overall decreased only slightly, from about 40,000 to about 38,000, in the narrower period 1981-1985 (IOM, 1990).

Even those who become interested in plants can hardly be encouraged when little graduate or postdoctoral support is available for plant sciences. Good researchers find their opportunities to train the next generation of plant scientists limited by the relatively small size and duration of the support of graduate and postdoctoral students.

The downward spiral (Figure 1 in Chapter 1) thus begins with a lack of funding, which reduces competitiveness and the number of plant scientists, and drives them into other fields, out of academia, or out of science. The spiral thus reduces the numbers of graduate and postdoctoral students, impoverishing the field. Reduction in the number of plant researchers and in the size of awards generates fewer important discoveries and causes the plant sciences to lag further behind other disciplines. This reduces the competitiveness of plant relative to animal projects and continues the cycle of reduced funding in plant biology. Ironically, the danger to the future of basic plant-science research is greatest now, when opportunities for science in general are greatest. Thus, fields for which funding is available will take advantage of the new breakthroughs and will progress at the expense of fields that are inadequately supported. Remedial action must be taken to correct the downward spiral.

Our analysis leads us to the following specific conclusions:

- Plant-biology research is not keeping pace with research in other fields of biology for several reasons:
  - Access to funding is limited.
  - There is no comprehensive system to support training and competitively awarded research grants.
  - The available grants are small and short term.
  - Few large research laboratories are performing forefront research using plant systems.
- There is insufficient basic plant biology in the core biology curricula of many universities and colleges.
- The amount of money available for the support of basic plant-biology research is inadequate relative to existing needs and opportunities for research and relative to support of other life-science programs.
- Federal support of plant biology is fragmented among many agencies.
- Most of the funds available to plant biology are targeted to the solution of immediate problems rather than to basic research. These funds often are not available broadly but are directed to scientists at specific institutions.

- There is no comprehensive program for financial support for graduate students and postdoctoral fellows in plant biology or for research support for trainees and faculty in plant biology.

If the problems are not remedied, plant biology in the U.S. can be expected to fall farther behind other sciences. The lack of emphasis on the teaching of plant sciences could worsen. Future practical applications–in fields as diverse as agriculture, nutrition, renewable energy, rangeland management, pharmacology, and management of the global environment–depend on our basic understanding of plants, so the failure to support plant-biology research and training will inhibit solution of these practical problems. The United States may thus become more dependent on scientific advances made abroad to support agriculture, one of its major industries and a major export earner in the U.S. economy.

The committee notes that the NIH system of comprehensive support for basic biologic research has been successful; that its elements are applicable to the problems facing the plant sciences; and that a program in basic plant sciences, constructed on the NIH model, would support research and teaching, and concomitantly would improve the competitiveness of U.S. plant science.

The NIH paradigm has four main features:

- NIH supports a diverse program of competitive grants for investigator-initiated extramural research at private and public universities and research institutions. Research grants are awarded to scientists who have applied to carry out their own projects within areas broadly defined by government policy.
- NIH's study section system is a critical element of its approach to the evaluation of research proposals. Study sections are composed of knowledgeable and objective scientists who review and evaluate ideas in grant

applications regardless of the projects' immediate applicability.

- One feature of the NIH health sciences program has been its support of graduate and postgraduate training and its awards to junior faculty. NIH has three early-career programs:
  - Graduate students are supported by training grants awarded to institutions with the strongest faculties and curricula, as judged by competition. An institution selects its graduate students, and the grant pays stipends and tuition costs for three to five years.
  - Postdoctoral fellows are supported by individual grants awarded by competition to scientists nearing completion of their graduate studies who have applied to carry out postdoctoral research. They are generally three-year awards.
  - Junior investigators are supported by individual, competitive grants awarded to junior faculty to defray their salaries and some part of their research costs for five years. Their universities release them from some teaching obligations for that period. The awards permit young scientists to redirect their interests and to spend much of their time doing research at the start of their careers.
- NIH has been the source of hundreds of millions of dollars for equipment and major construction at universities (although the amounts awarded have decreased sharply in recent years). The program has been of inestimable value in increasing the pace of research at recipient institutions. NIH's grants for research, training, other infrastructural elements, such as instruments, facilities, symposia, meetings, and public information has resulted in the construction of a comprehensive and complete system of support for biomedical research.

Individual investigators, small groups of investigators, many of them university based, make up the backbone of American science. It was enlightened support over the past decades of that group of individuals that has given the United States a research and technology enterprise that is the envy of the world. (D. Allan Bromley, Science Advisor to the President, in a speech at the National Academy of Sciences, June 27, 1990)

4

RECOMMENDATIONS

The members of the Committee on Plant Sciences believe it is time to apply to the plant sciences the lessons learned from the support of biomedical research and training. The committee recommends the establishment of a National Institute of Plant Biology (NIPB) with a comprehensive program that engages all of the federal agencies that support plant biology. NIPB would be organized in the U.S. Department of Agriculture (USDA). The institute would be based on the principle of competitively awarded basic research and training grants in plant biology and its philosophy and practice would be patterned after the National Institute of General Medical Sciences (NIGMS) of the National Institutes of Health (NIH).

The term "program" refers to the framework proposed by the committee, including the establishment of, and leadership role to be played by, the institute in USDA; the vital continued commitment and participation of other agencies in support of plant biology research and training in cooperation with NIPB; a study section system; and the kind of support and amount of funding proposed by the committee that are essential to the program's success.

MANAGEMENT OF THE PLANT-BIOLOGY PROGRAM

The success of the proposed plant biology program will depend on its meeting the following criteria:

- The program should be dedicated to the study of plant biology as a basic science. It should not be a mission-oriented program aimed at solving specific practical problems.
- The program should encompass a comprehensive system of extramural research and training to include pre- and postdoctoral fellowships, training grants for graduate students, grants for the purchase and upkeep of instrumentation, and financial support for meetings. The system of grants should support the highest quality research in nonprofit institutions.
- The program should be patterned in the detail of its technique and philosophy after NIGMS.
- The program should support high-quality research being done by plant biologists in nonprofit institutions. Communication between plant scientists and researchers in other disciplines should be encouraged.
- The program should provide grants and fellowships in sufficient number and amount of award to attract and retain the best scientists.
- The program should be administered by an agency committed to the above standards.

RECOMMENDATION 1

Because the selection of an agency to lead a coordinated effort to promote plant biology within the federal system is critical, the committee weighed a range of options. Initially, its members focused on identifying a single agency that would have almost exclusive responsibility for the entire program in plant biology. Several options that named a single agency to administer the program were rejected for failure to satisfy a critical scientific, managerial, or political need. The committee eventually concluded that it must define a multiagency effort with one agency taking a decided leadership role. Only in this way could the range of societal and scientific needs in medicine, agriculture, the environment, and energy be addressed.

A National Institute of Plant Biology (NIPB) should be established in USDA under the direct oversight of the assistant secretary of agriculture for science and education. NIPB should be responsible for leading a coordinated federal plant-biology program that intimately involves other federal agencies that support plant-biology research and training.

The recommendation that USDA should be the lead agency to assume broad responsibility for the support of plant sciences (in concert with other agencies) is made with full awareness of the historic mission of, and current practice at, USDA. USDA's mission and the largest part of its funds traditionally have been dedicated to formula support of research in designated land grant schools and in its intramural agricultural stations. The formula funding that served U.S. agriculture successfully for the first half of this century has not provided a mechanism to keep abreast of the spectacular advances in modern biology, and support of training has not been a primary objective of USDA funding. Plant-biology research is a broad endeavor and USDA's agricultural mission is too narrow to encompass all the fundamental plant biology we believe should be included in the program. Political and commercial influence on the department's decisions and a tendency to overmanage the research process (NRC, 1972) have impeded the development of fundamental research programs.

Two major factors led us to recommend USDA to establish NIPB and take leadership of the federal plant biology program. First, the attempts since 1978 to broaden the USDA base of support for basic agricultural research through a program of competitive grants indicates that the almost exclusive concentration on formula grants that characterized USDA is changing. The fiscal year 1992 initiative to enlarge USDA's small program of extramural grants brings a welcome competitive process for research support to a few segments of plant biology related to agriculture. When the initiative is fully funded at $500 million, $125 million is proposed for expenditure on plant biology. Although our recommendation builds on the foundation of the National Research Initiative Competitive Grants Program (NRICGP), it goes beyond that program in proposing that the plant systems portion become the core of NIPB. In addition, NIPB would lead the coordination of efforts in plant biology sponsored by other agencies.

The second factor influencing our recommendation is that of all the agencies with potential to lead the plant program, USDA's mission encompasses the broadest range of scientific and applied interests; it includes research on plants, forestry, nutrition, rangelands, and the ecological relationships of plants to other biotic and nonbiotic systems. NIH and the Department of Energy (DoE) have been sympathetic in support of several aspects of plant biology, but neither has the breadth of interest in plants to make it a natural home for the new institute.

Implementation of our proposal would require that USDA effect major changes in its philosophy of research, its operational patterns, and its relationship to Congress and the scientific community. It will need to

- Plan beyond the design drawn for NRICGP and its proposed five-year funding strategy.

- Support evolution of NRICGP and its *competitive* grants program.
- Focus on the support of *fundamental* plant biology.
- Insulate the new institute from political and commercial pressures.
- Avoid over managing the scientific research process.
- Demonstrate increased leadership in coordinating its work with that of other agencies.
- Develop department-supported training programs and encourage training programs at other agencies.
- Organize study sections that use the expertise of the entire scientific community by reaching outside the government.
- Organize NIPB to ensure its high visibility, stature, and independence within the federal government.

The National Science Foundation (NSF) has a scope of interests that overlaps that of USDA and historically has provided more financial support for competitively awarded, investigator-initiated plant-biology research than has USDA. However, we believe that NSF's multitude of other interests would impede its serving as the lead agency for the new program. Should USDA prove unwilling to fulfill the role we have described for it, NSF should be assigned the task of leading the program, for NSF has clearly demonstrated its dedication to the support of fundamental research based on competitively awarded, investigator-initiated grants.

RECOMMENDATION 2

NSF, DoE, NIH, and the National Aeronautics and Space Administration (NASA) have provided valuable support for plant biology research, and their continued financial support at increased levels will be required to fulfill the objectives of the

USDA-led program. Taken as a group, the agencies have missions that encompass all aspects of a complete plant-biology program, from molecular biology to ecosystem research. NIH and NSF could provide the training grants and fellowships that are essential to the development of a larger number of plant scientists. However, we urge USDA to explore the possibilities of developing training programs of the size we propose. For our plan to succeed, all agencies, including USDA, will need to increase the amount awarded in individual research grants.

All agencies that currently support plant-biology research and training should maintain and increase their commitment in cooperation with NIPB and USDA.

The Office of Science and Technology Policy (OSTP) is responsible for coordinating interagency research. It discharges this responsibility increasingly through the formation of Federal Coordinating Councils for Science, Engineering and Technology (FCCSET). Creation of an OSTP FCCSET committee on plant research that is chaired by a USDA official might be an effective means for coordinating the research. It should be noted that FCCSETs often are comprised of department level members who do not manage specific programs directly. On the other hand, for some years an interagency coordinating committee, made up of persons closely affiliated with agency plant-biology programs, has worked well, for example, to organize interagency funding of large-scale centers. It might be advantageous for OSTP to seek ways to make the best use of both a FCCSET and the existing committee in its efforts to coordinate plant-biology research.

If the challenge is successfully met, the establishment of NIPB would be another step in an important progression. The first step was the establishment of USDA's competitive grants program; the second was the expansion of that program under the National Research Initiative. Potentially, other parts of the

National Research Initiative Competitive Grants Program, such as the program in animal health, could become institutes as well. Eventually, USDA could follow the model of NIH in the Department of Health and Human Services for support of extramural and intramural research, training, and the infrastructural elements of sciences relevant to its mission.

ELEMENTS OF THE NIPB PROGRAM

NIPB would manage a comprehensive program of support for research, training, facilities, and scientific communications. Awards would be made by unambiguously competitive, peer-reviewed procedures open to all scientists. NIPB would coordinate the existing support from several government agencies, and, with increases in these agencies' existing competitive grants programs, would give the nation the infrastructure for plant biology that it now lacks.

We underscore the pivotal importance of competitive, peer-reviewed procedures. In the 45 years since the beginning of large-scale federal support of science, the strategies used by the various federal agencies to fund scientific research in support of societal goals have constituted an experiment. NIH and NSF have based funding decisions on competitive procedures designed to recognize individual merit; USDA has based funding decisions on institutional, political, and historical considerations that do not preclude but that also do not necessarily reward or reinforce individual merit. The committee concludes that the results of the experiment are clear. The philosophy, mechanisms, and strategy used by NIH and NSF to support basic research and its applications have advanced science of the highest quality, attracted the best young scientists to careers in research and teaching, and provided a stream of discoveries that has been rapid and highly beneficial to society. The success of the NIH and NSF grant programs has engendered their enthusiastic and generous support by Congress and successive administrations.

The projected program of NIPB should include the following program components and management features.

Individual Research Grants

The core of NIPB's program should be competitively awarded, investigator-initiated grants to researchers in any institution of higher education or advanced research. The essential criterion for award of a grant should be scientific merit.

Grants generally should be for a five-year period and have an average total cost per grant of $170,000 per year. This is the same as the average NIGMS grant. There should be adequate provision for institutional overhead and administrative expenses.

Peer review of grant applications should be conducted by study sections of qualified reviewers. The scope of the program should be carefully defined in the course of further study but the following subjects are cited as examples:

- *Subcellular processes*, including biochemistry, photochemistry, organelle structure and function, gene and chromosome structure, genome organization, mutagenesis and DNA repair, and gene expression and regulation.
- *Cellular processes*, including developmental biology and developmental genetics, signal transduction, cell-to-cell communication, cell division and growth, photosynthesis, and intercellular transport of water and nutrients.
- *Organismal processes*, including growth and reproductive biology, structure and function of plant organs, responses to the environment at the supercellular level, and nutrient and water transport in the whole plant.

- *Population and species processes*, including areas such as ecology, population biology and genetics, systematics, and issues of biodiversity.
- *Plant interactions with the biotic and abiotic environment*, including nitrogen fixation, interactions with beneficial microorganisms, pathogenesis, the genetics and molecular biology of plant defense and stress responses, and community ecology.

Competitive Postdoctoral Training Awards

The proposed program would support postdoctoral training in basic plant biology, because postdoctoral experience is necessary to complete the training of our most promising researchers. An attractive program will bring additional postdoctoral fellows to plant biology from other predoctoral disciplines.

Awards would be based on review by qualified panels of scientists. Applications would be filed either before or after an applicant's receipt of the Ph.D. degree. The nature of the host laboratories and their location in the United States or abroad would not be restricted.

Predoctoral Training Awards

The proposed program would support training grants similar to those funded by NIH. These would support a number of students, and the grants would be awarded to the institutions' departments. Individual predoctoral fellowships, similar to those sponsored by NSF also would be awarded.

Departmental Training Grants

The program is projected to provide support to build strong departments of plant science and to strengthen the programs of other departments that include plant research and training. By the year 2000 a total of 34, five-year-long departmental grants is proposed (each renewable for five years). Participation by 15 students per program is projected, although the number would vary. To be attractive, the stipends would be comparable to those for other natural sciences. Funds would be provided to the universities to cover tuition, and supply allowances would be granted to the laboratories of the students' supervisors.

Applications for the grants would be submitted by departments, and the competitively awarded grants would provide steady funding for outstanding training programs.

Individual Predoctoral Fellowships

The program would provide individual fellowships to highly qualified predoctoral candidates. Candidates would apply either in the senior year of undergraduate study or in the first year of graduate study.

The program would provide four-year awards, and a total of 1,500 fellows would be supported when the program is fully implemented. Stipends would be competitive with those provided to students in other natural sciences and somewhat above those for departmental awards. Funds would be awarded to the universities for tuition and for supplies in individual laboratories.

A recipient would be allowed to choose a host research laboratory. This would provide additional support to superior programs and would stimulate competition among schools for the participating fellows.

Summer Undergraduate Training

The program would include support of summer undergraduate research. When fully implemented, it would support up to three students in the laboratories of scientists who have been recognized through the award of research grants. By the year 2000, summer research experience would be provided to about 1,500 students.

Career Training and Redevelopment

The program would provide retraining and continuing education for faculty members and facilitate communication among investigators at different institutions. The first component of the program would provide funding for sabbatical leaves for up to one year for 100 persons in the year 2000.

The second component would provide salary for faculty from predominantly teaching institutions or from institutions with few graduate students to work in active research laboratories, generally during the summer. It would support three-month-long summer fellowships for 50 persons each year. Requests for support would be submitted by individuals, and the fellowships would be awarded competitively based on peer review.

Facilities and Equipment

The program would provide support for instrumentation and facilities. Applications would come from departments, and a grant pool of $10 million per year would be awarded competitively. This would provide for individual and shared facilities in departments with competitive plant-biology funding and would provide funds for the purchase of new equipment and facilities and for replacement of obsolete equipment and facilities.

Scientific Communications

The program would help support plant-biology symposia by providing partial funding for travel and subsistence of participants at 20 scientific meetings each year. Other innovative ways to foster reciprocal scientific communication among the plant sciences and other fields should be encouraged. For example, computer networks, data base and germplasm information and materials sharing, and teleconferencing would be supported. Support for expansion of existing journals to include the plant sciences would be considered.

Figure 4 shows the relationships among the components of the program.

RECOMMENDATION 3

The committee's members believe there should be special provision for continuing, independent advice and periodic evaluation.

An independent group of non-government scientists should be formed to provide continuing advice to the USDA assistant secretary for science and education and to the officials of cooperating agencies concerning NIPB's operation and direction and to oversee the parallel efforts by other agencies.

Moreover, after five years an independent group should examine and evaluate the progress of all agencies in implementing the recommendations contained in this report.

Figure 4 Framework of plant-biology research and training program.

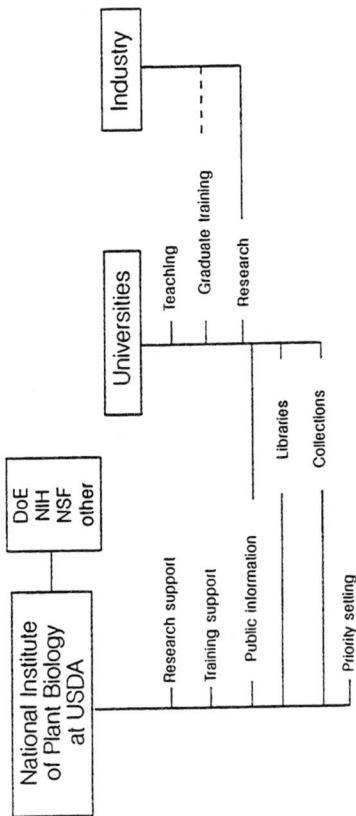

It is the usual practice at federal research agencies to form an advisory council. The Independent Advisory Group (IAG) we recommend follows that pattern. The group's first priority would be to give advice on and review the design of an action plan drafted by USDA scientists and policy makers, representatives of the academic research and training community, and the cooperating federal agencies. The action plan would describe the strategy and detail the organization, structure, and schedule for establishing the institute and implementing its program. Thereafter, IAG would serve as a scientific board of advisors to the assistant secretary overseeing progress toward the goals described in the action plan and suggesting corrections and additions to the plan as dictated by events and experience.

At the end of five years, a separately constituted, independent, nongovernment group would review the program's performance comprehensively and recommend changes.

COST ELEMENTS AND SIZE OF THE PROGRAM

The program described here represents the combined support and efforts of several federal agencies, with NIPB serving as the lead in coordinating the effort. The size and cost recommended for the program are predicated on the following reasoning:

An NSF survey (NSF, 1990b) reported that there are about 4,500 full-time plant-biology faculty in academic departments. Seventy-nine percent of these faculty members (about 3,600) train graduate students. We use training of graduate students as a surrogate determinant for estimating the number of active research faculty. We estimate that 20% of the 3,600 plant biologists would not be part of a grant applicant pool because they already receive support from other sources or because they would not compete well for funding. Thus, the estimated base number of current scientists who would be part of the applicant pool is about 3000.

Over the course of the nine years shown in Table 3 to the year 2000, several considerations described below could increase the numbers in the applicant pool. If our proposed program were implemented and adequate funds were provided, young scientists would be encouraged to enter plant-biology research careers and some active scientists would have an incentive to shift their interest to the study of plant models that often offer advantages over animal or microbial models. The projected training programs would augment the skilled cohort of scientists in the applicant pool. The NSF survey predicts a potential immediate increase in the applicant pool because there are 276 unfilled faculty positions in academic plant-biology programs. Furthermore, departments of biology whose hiring practices have been influenced by considerations of the "fundability" of candidates would be encouraged to seek plant biologists to balance their programs. There is evidence from a directly relevant program that the increased availability of funds increases the numbers of applicants. Applications for plant-systems research support received by NRICGP increased from 1,287 in 1990 to 1,793 in 1991.

We estimate conservatively that the number in the applicant pool would reach 6,000 by the year 2000. Our suggested program is aimed at providing a success rate (percentage of total applicants that receive awards) of 40%. This would provide for healthy competition and support of appropriate numbers of superior applicants.

About 1,350 awards (individuals could have several awards) currently are made by the agencies and programs listed in Table 2 (see Chapter 2). Assuming that the applicant pool is now about 3,000 individuals, the success rate among current applicants is about 40%. For those who are successful in obtaining support, the major issues are the size and duration of grants and the lack of funds to support training, career development, and facilities.

Table 3   Number of awards and financial support (in thousands of 1991 dollars) for the plant biology program

Program element	1992	1993	1994	1995	1996	1997	1998	1999	2000
Individual Research grants (at $170,000/award)	1500 / $255,000	1700 / $289,000	1900 / $323,000	2000 / $340,000	2100 / $357,000	2200 / $374,000	2300 / $391,000	2400 / $408,000	2500 / $425,000
Postdoctoral fellowships (at $35,000/award)	75 / $2,625	150 / $5,250	200 / $7,000	250 / $8,750	300 / $10,500	350 / $12,250	400 / $14,000	450 / $15,750	500 / $17,500
Predoctoral (departmental) training grants (at $360,000/award) No. trainees	10 / $3,600 / 150	15 / $5,400 / 225	20 / $7,200 / 300	24 / $8,640 / 360	26 / $9,360 / 390	28 / $10,080 / 420	30 / $10,800 / 450	32 / $11,520 / 480	34 / $12,240 / 510
Individual predoctoral fellowships (at $29,000/award)	200 / $5,800	400 / $11,600	600 / $17,400	800 / $23,200	1000 / $29,000	1200 / $34,800	1300 / $37,700	1400 / $40,600	1500 / $43,500
Summer undergraduate training program (at $4,000/award)	200 / $800	400 / $1,600	600 / $2,400	800 / $3,200	1000 / $4,000	1200 / $4,800	1300 / $5,200	1400 / $5,600	1500 / $6,000
Sabbatical leaves (at $50,000)	25 / $1,250	50 / $2,500	100 / $5,000	100 / $5,000	100 / $5,000	100 / $5,000	100 / $5,000	100 / $5,000	100 / $5,000
Summer fellowships (at $15,000)	25 / $375	50 / $750	50 / $750	50 / $750	50 / $750	50 / $750	50 / $750	50 / $750	50 / $750
Facility support (at $1,000,000)	10 / $10,000	10 / $10,000	10 / $10,000	10 / $10,000	10 / $10,000	10 / $10,000	10 / $10,000	10 / $10,000	10 / $10,000
Meeting support (at $25,000)	20 / $500	20 / $500	20 / $500	20 / $500	20 / $500	20 / $500	20 / $500	20 / $500	20 / $500
Total support	$279,950	$326,600	$373,250	$400,040	$426,110	$452,180	$474,950	$497,720	$520,490

We propose support for training sufficient to encourage students to study plant biology and to create a pool of new plant biologists for academia, the government, and industry. The NSF survey reports that in 1988-1989, there were 7,317 graduate students and about 1,120 postdoctoral fellows in this field. Twenty-one percent of the graduate students and 53% of the postdoctoral fellows were supported by federal research grants; 4% of the graduate students and 7% of the postdoctoral trainees were on federal fellowships. Graduate students also are supported by other sources, including institutions (28%), state governments (15%), and personal funds (11%). Other sources of support for postdoctoral fellows include state governments (11%), foreign governments (7%), and industry (7%). Our projected program would provide for individual fellowships and departmental training grants *in addition to the already existing support from other sources, including from research grants.* The number of trainees will increase if funding is available, thus reversing a trend of decreasing numbers of graduate students in plant-biology programs.

We believe that several support mechanisms for trainees will be needed to achieve the target of a 50% increase by the year 2000. Funding opportunities for trainees would be increased by larger research grants. The introduction of major training grants would encourage highly qualified trainees to enter the field of plant biology. In the year 2000, such grants could support about 10,500 graduate students and 1,600 postdoctoral fellows. These estimates are based on a projection of 6% annual growth in the number of trainees from the year 1988-1989. Eventually, about 4,250 graduate students would be supported by the combination of departmental training grants (750), individual predoctoral fellowships (1,500), and research grants (2,000). Using the same assumptions, about 1,300 of the 1,600 postdoctoral researchers would be supported by a combination of 500 fellowships and 800 research grants.

Table 3 shows the increasing number of awards from 1992 to 2000 that would fulfill our estimate of minimal needs for research and training support. The 1,500 grants shown for the first year approximate the grants that would be active at that time; approximately 1,300 are now active. The first-year sum encompasses approximately $150 million already in the budgets of the agencies listed in Table 2. Most of the increment arises from our proposal that the size of grants be increased substantially and that training and other program elements be implemented. Incremental growth in the research grant category as well as in other categories is based on conservative estimates of growth. For example, 10 departments would receive training grants in the first year to support about 15 predoctoral students each. The number of departments with training grants is projected to increase rapidly for the first several years and then level off as the new programs mature.

The progressive increase in the number of awards in the period until the year 2000 shown in Table 3 is the first phase of the program, and it provides a period to test the effectiveness of the program and to adjust it as needed. We anticipate that the program will continue to grow after the year 2000 beyond the figures shown for that year.

We consider that the program presented here constitutes the minimum effort necessary to ensure U.S. leadership in plant-biology research into the next century.

LITERATURE CITED

American Association for the Advancement of Science, Intersociety Working Group. AAAS Report XV: Research and Development FY 1991. Washington, D.C.: AAAS, 1990. 330 pp.

Institute of Medicine, Committee on Policies for Allocating Health Sciences Research Funds. Funding Health Sciences Research: A Strategy to Restore Balance. Edited by F. E. Bloom and M. A. Randolph. Washington, D.C.: National Academy Press, 1990. 255 pp.

Kerr, N. A. The Legacy: A Centennial History of the State Agricultural Experiment Stations 1887-1987. Columbia, MO.: Missouri Agricultural Experiment Station, 1987. 318 pp.

National Association of State Universities and Land-Grant Colleges, Committee on Biotechnology. By R. H. Biggs, L. W. Moore, and R. A. Dreher. Instrumentation and Equipment Survey for Agricultural Biotechnology: Final Report. November 1989. [No publisher] 85 pp.

National Association of State Universities and Land-Grant Colleges, Higher Education Statistics Committee. Fall 1989 Enrollment in NASULGC Colleges of Agriculture. [No publisher.] 1990. 23 pp.

National Research Council, Report of the Committee on Research Advisory to the U.S. Department of Agriculture. Washington, D.C.: National Academy of Sciences, 1972. 463 pp.

National Research Council, Investing in Research: A Proposal to Strengthen the Agricultural, Food, and Environmental System. Washington, D.C.: National Academy Press, 1989a. 155 pp.

National Research Council, Opportunities in Biology. Washington, D.C.: National Academy Press, 1989b. 448 pp.

National Research Council, Forestry Research: A Mandate for Change. Washington, D.C.: National Academy Press, 1990. 84 pp.

National Science Foundation. A Long-range Plan for the Multinational Coordinated *Arabidopsis thaliana* Genome Research Project. Washington, D.C.: National Science Foundation, 1990a. 14 pp. (NSF 90-80)

National Science Foundation, Directorate for Biological, Behavioral, and Social Sciences. Plant Biology Personnel and Training at Doctorate-Granting Institutions. Prepared by B. Chaney, E. Farris, and P. White. Higher Education Surveys Report 13. Washington, D.C.: National Science Foundation, 1990b. 73 pp.

National Science Foundation, Federal Funds for Research and Development: Fiscal Years 1988, 1989, and 1990. Volume 38. Detailed Statistical Tables. Washington, D.C.: National Science Foundation, 1990c (NSF 90-306.)

U. S. Department of Agriculture, Cooperative State Research Service. Food and Agriculture Competitively Awarded Research and Education Grants: Fiscal Year 1989. Washington, D.C.: U.S. Government Printing Office, 1990a. 211 pp. (261-494/20080.)

U. S. Department of Agriculture, Cooperative State Research Service. Fostering Fundamental Research for the Future. By S. J. Rockey et al. Washington, D.C.: USDA, 1989. 18 pp.

U. S. Department of Agriculture, Program Plan for the National Initiative for Research on Agriculture, Food and Environment. Washington, D.C.: USDA, 1990b. 25 pp.

U.S. Department of Health and Human Services, Public Health Service. Basic Data Relating to the National Institutes of Health, NIH Data Book 1991. Washington, D.C.: DHHS, 1991.

INFORMATION ON COMMITTEE MEMBERS

Robert Goodman (chairman) is a professor at the University of Wisconsin–Madison. He was previously executive vice-president for research and development at Calgene, Inc., and prior to that a professor at the University of Illinois at Urbana-Champaign. His research is in the genetics of plant defense responses and plant virology. He is a member of the NRC Board on Agriculture and was a scholar-in-residence at the research council. He served on the committees of the Board on Agriculture that wrote *Investing in Research* and *Alternative Agriculture.*

John Axtell is a professor of genetics, at Purdue University, Lafayette, Indiana. His research focuses on plant breeding. He is a member of the National Academy of Sciences

Frederick A. Bliss is a professor of pomology and holds the Will W. Lester Endowed Chair at the University of California, Davis. He is a plant geneticist with interests in plant breeding and development. He was a member of the National Plant Genetic Resources Board and has conducted research in Nigeria, Latin America, and Germany.

Winslow R. Briggs is the director of the Department of Plant Biology, Carnegie Institution of Washington, Stanford, California. He is a plant physiologist with research interests in plant growth and development and in the physiology, biochemistry, and molecular biology of photomorphogenesis. He received the Alexander von Humboldt U.S. Senior Scientist Award and is a past president of the American Society for Plant Physiology. He is a member of the National Academy of Sciences and the American Academy of Arts and Sciences.

65

Donald D. Brown is the director of the Department of Embryology, Carnegie Institute of Washington in Baltimore, Maryland. His research focuses on animal developmental biology. He has served on numerous scientific committees and initiated and manages a fellowship program for outstanding researchers. He is a recipient of the U.S. Steel Foundation Award for Molecular Biology and other honors and is a member of the National Academy of Sciences.

Michael T. Clegg is a professor of botany and genetics, University of California, Riverside. His research interests are in population genetics and evolution. He is a member of the National Academy of Sciences.

Jeffrey J. Doyle is an associate professor at the L. H. Bailey Hortorium, Cornell University, Ithaca, New York, where he performs research on plant systematics and molecular evolution.

James R. Ehleringer is a professor in the Department of Biology, University of Utah. His research is in plant physiology and plant ecology, with an emphasis on linkages between these areas. He has served on major advisory panels for federal agencies and on editorial boards; he is currently editor-in-chief of *Oecologia*.

Gerald R. Fink is the American Cancer Society Genetics Professor and director of the Whitehead Institute for Biomedical Research, Massachusetts Institute of Technology, Cambridge, Massachusetts. His research interest is in plant and fungal molecular genetics. He is a member of the National Academy of Sciences and a recipient of the U.S. Steel Foundation Award for Molecular Biology and other honors.

Robert B. Horsch is manager of crop transformation for Monsanto Agriculture Company, St. Louis, Missouri, where he performs plant biotechnology research. He is an adjunct professor in the Department of Biology, Washington University, St. Louis, Missouri, and serves on the editorial boards of several plant-science journals. He has organized courses in plant molecular biology at Cold Spring Harbor Laboratories.

Elliot M. Meyerowitz is a professor of biology at the California Institute of Technology, Pasadena, California. His research interests are in genetics, plant development, and molecular genetics using both animal and plant models.

Paul G. Risser is provost and vice-president for academic affairs, University of New Mexico, Albuquerque. He is a past president of the American Ecological Society and has served on National Research Council committees on global change and other environmental topics. He is the current chairman of the Board on Environmental Studies and Toxicology.

Susan R. Wessler is an associate professor of botany, University of Georgia, Athens. Her research involves plant molecular genetics. She has served on a number of National Institutes of Health and National Science Foundation panels and is coeditor of *The Plant Cell*.